CATERPILLAR
PHOTO GALLERY

Iconografix seeks collections of archival photographs for reproduction in future books. We require a minimum of 120 photographs per subject in our Photo Archive Series and Photo Album Series, and a minimum of 500 photographs per subject in our Photo Gallery Series. We prefer subjects narrow in focus, i.e., a specific model, or manufacturer, railroad, racing venue, etc. Photographs must be of high-quality, suited to reproduction in an 8x10-inch format. We willingly pay for the use of photographs.

If you own or know of such a collection, please contact: The Publisher, Iconografix, PO Box 446, Hudson, Wisconsin 54016, USA.

CATERPILLAR
PHOTO GALLERY

Edited by P.A. Letourneau

Iconografix
Photo Gallery Series

Iconografix
PO Box 446
Hudson, Wisconsin 54016 USA

Text Copyright © 1997

All rights reserved. No part of this work may be reproduced or used in any form by any means... graphic, electronic, or mechanical, including photocopying, recording, taping, or any other information storage and retrieval system... without written permission of the publisher.

We acknowledge that certain words, such as model names and designations, mentioned herein are the property of the trademark holder. We use them for purposes of identification only. This is not an official publication.

Iconografix books are offered at a discount when sold in quantity for promotional use. Businesses or organizations seeking details should write to the Marketing Department, Iconografix, at the above address.

Library of Congress Card Number 97-70622

ISBN 1-882256-70-0

97 98 99 00 01 02 03 5 4 3 2 1

Printed in the United States of America

Preface

Few companies have gained as significant a share of their respective worldwide product markets as has Caterpillar, Inc. Fewer yet have earned the same degree of name recognition, or gained a comparable reputation for product excellence.

In achieving such success, Caterpillar has also gained the admiration of tens of thousands of people who hold a fascination for crawler tractors. It is just for this group of enthusiasts that Iconografix has published *Caterpillar Photo Gallery*, a collection of over 500 archival photographs of track-laying tractors built by Caterpillar, Inc. and its two principal predecessor companies, Holt Manufacturing Company and C.L. Best Tractor Company. Beginning with Benjamin Holt's first experimental steam crawler and concluding with the mighty diesel-powered Cat D9G, more than 60 "Tracklayer" and "Caterpillar" models are featured at work in operations that include earthmoving, farming, logging, military support, mining, and more.

Caterpillar Photo Gallery is organized by model and presented in near-chronological order. Holt tractors built between 1904 and 1925 appear first; followed by Best tractors built between 1912 and 1925; concluding with machines built after the April 1925 merger that created Caterpillar Tractor Co. The majority of photographs were taken for publicity purposes, and were used in sales literature and/or were freely distributed with captions to trade magazines, newspapers, and dealers, a practice less common in today's world of animated advertising and communication. For the most part, the captions that accompany the photographs are condensed from those written by the respective companies' publicity departments. Sometimes the companies noted the date of a photograph and the location at which it was taken. Just as often, they did not. Consequently, not all captions include dates and locations. Not every model of crawler tractor built in the history of the three companies is included in the book, as, unfortunately, only a limited number of original factory photographs have been preserved. Nonetheless, *Caterpillar Photo Gallery* is the most comprehensive collection of Caterpillar tractor photographs ever published. It is our hope that readers will find the book a fascinating and valuable addition to their libraries.

We wish to acknowledge the contributions of the Higgins Collection, Shields Library, University of California, Davis, and the Caterpillar Inc. Corporate Archives. These images would have been lost long ago, if it were not for their commitments to preserving Caterpillar's history. We also thank the Antique Caterpillar Machinery Owners Club, 10816 Monitor-McKee Road NE, Woodburn, Oregon 97071, USA, certain members of which reviewed and offered corrections to the book's manuscript. Anyone with an interest in Caterpillar machinery is encouraged to write the club for membership information.

Highlights in the History of Caterpillar Tractors 1904-1955

1904
Holt Manufacturing Company, Stockton, California, builds its first experimental track-type tractor.

1906
Holt begins production of steam crawlers, and organizes Aurora Engine Company to build gasoline engines.

1908
Holt begins production of gasoline-powered crawlers.

1909
Holt establishes production in Peoria, Illinois.

1910
C.L. Best establishes C.L. Best Gas Traction Company, Elmhurst, California, and begins production of gasoline wheel tractors.

1912
Best introduces "C.L.B." 75 H.P. Tracklayer, the company's first crawler tractor.

1918
Holt introduces T-11 and T-29, later designated 5 Ton. Production ends 1926.

Holt introduces 10 Ton. Production ends 1926.

1919
Best introduces Sixty. Production ends 1931.

1921
Holt introduces T-35, later designated 2 Ton. Production ends 1928.

Best introduces Thirty. Production ends 1931.

1925
Best and Holt merge as Caterpillar Tractor Company.

1926
Work begins on adapting diesel engine for use in crawler.

1927
Twenty introduced. Production ends 1932.

1929
Ten introduced. Production ends 1932.

Fifteen introduced. Production ends 1932.

1931
Diesel Sixty, first production diesel-powered tractor, introduced; later designated Diesel Sixty-Five. Production ends 1932.

Twenty-Five introduced. Production ends 1933.

Gas Fifty introduced. Production ends 1937.

1931 (continued)
Gas Sixty-Five introduced. Production ends 1932.

1932
Gas Thirty-Five introduced. Production ends 1934.

Twenty (8C-Series) introduced. Production ends 1933.

High Clearance Fifteen introduced. Production ends 1933.

Fifteen (7C-Series) "Small Fifteen" introduced. Production ends 1933.

1933
Twenty-Eight introduced. Production ends 1935.

Diesel Thirty-Five introduced. Production ends 1934.

Diesel Fifty introduced. Became RD-7 in 1935; D7 in 1938.

Gas Seventy introduced. Production ends 1937.

Diesel Seventy introduced. Production ends 1933.

Diesel Seventy-Five introduced. Became RD-8 in 1935; D8 in 1938.

1934
Twenty-Two introduced. Production ends 1939.

R3 introduced. Production ends 1935.

R5 introduced. Production ends 1940.

Gas Forty introduced. Production ends 1936.

Diesel Forty introduced. Became RD-6 in 1935; D6 in 1938.

1935
R2 introduced. Production ends 1942.

Thirty (6G Series) introduced. Became R-4 in 1935. Production ends 1944.

1936
RD-4 introduced. Became D4 in 1939.

1938
D2 introduced.

1939
D5 introduced. Production ends 1939.

1941
R6 introduced. Production ends 1941.

1955
D9 introduced.

Auxiliary rear wheels and front "wooden barrel" wheel were Benjamin Holt's earliest solution to preventing his tractors from sinking into the soft soil of California's San Joaquin Valley.

A Holt steam tractor was a heavy machine. If the auxiliary wheels, as fitted to the tractor at the top of this page, could not keep it afloat, then certainly wider wheels were the answer. This Holt steamer was fitted with 42-inch drive and 42-inch auxiliary wheels. There had to be a better solution...

In 1904, Holt began experiments with tracks. This steam unit, converted from wheels to tracks, became a prototype that tested Holt's "Caterpillar" principle.

The first gas Caterpillar, serial no. 1001, with a Holt harvester. Built in 1906, no. 1001 underwent extensive testing before no. 1002 was built in 1908.

The advent of the gasoline engine meant tractors of smaller size and lighter weight produced the horsepower equivalent of heavyweight steam units. Shown undergoing a factory-conducted test is no. 1002, the second gas Caterpillar tractor, built in 1908.

Holt Model 40, serial no. 1003, the first gas Caterpillar tractor sold. This was one of 28 tractors delivered to the city of Los Angeles between 1908 and 1910.

Rear view of an early Holt steam Caterpillar tractor with 9-foot track. Track shoes were made of wooden slats. Note the chain drives, circa 1908.

By 1909, Holt gasoline-powered Caterpillar tractors were in regular production. Here, two Model 40s are loaded for shipment to the Sacramento Valley Sugar Co., Hamilton City, California.

Holt steam Caterpillar no. 122 featured a 10-1/4 x 12-inch cylinder, and was equipped with 30-inch wide track, circa 1910.

A Model 40 pulling nine 14-inch plow bottoms on a 960-acre Oregon farm.

Freighting was a major market for early Caterpillar tractors. Here, two Model 40s freight construction materials across the Mojave Desert.

The construction of the Los Angeles Aqueduct was a major test of the early Caterpillar tractor. The results were a near disaster for Holt, as a final review indicated that horses would have done the same work for one-half the cost.

Holt Model 40-45 with road grader blade.

Holt Model 40-45 pumping water for irrigation, Lovelock, Nevada, 1910.

Holt Model 45, serial no. 1105, was delivered June 1910 to the C. Parker Holt Ranch. The 4-cylinder engine, rated at 45 brake horsepower, featured 6-1/2 x 8-inch bore and stroke.

A Holt wide-tread Model 45, serial no. 1178, featured improved steering, wood rim steering wheel, cloth top, four track rollers, and 6-1/2 x 8-inch engine.

Holt Model 60, serial no. 1226, one of ten Caterpillars shipped to Argentina in 1911. The Holt 60 featured a 7 x 8-inch, 4-cylinder engine rated at 60 brake horsepower.

Rear view of the Holt 60 reveals a belt pulley. This unit was built in Stockton, California. The Holt 60 was also built in Peoria, Illinois.

LET THE CATERPILLAR SOLVE YOUR LOGGING PROBLEMS

F OR OVERLAND FREIGHTING, where heavy loads are impractical with teams or ordinary types of wheel tractors, the HOLT CATERPILLAR has great advantages; it was designed for just such work.

It operates successfully over loose, wet or sandy soils—will not slip—will not mire, because of its great bearing surface; and its pulling power, with low cost of up-keep and operation, combine to make it THE IDEAL TRACTIVE POWER FOR THE LOGGING COUNTRY.

IT COSTS NOTHING WHEN NOT IN ACTUAL USE.
IT WILL SUCCESSFULLY REPLACE TEAMS
AND DO MORE WORK AT LESS EXPENSE.

FURNISHED IN TWO SIZES:
        45 rated horse power with 30 H. P. draw bar capacity
        60 rated horse power with 40 H. P. draw bar capacity

Catalog sent upon request.

The Greatest Tractive Power of the Age

MANUFACTURED BY

Holt Caterpillar Company

PEORIA, ILLINOIS

New York Office   - -   50 Church Street

H. B.   No. 50

Holt "Baby" 30, 1915.

Holt "Baby" 30, 1915.

Holt "Baby" 30 with disc and spring-tooth harrow, 1915.

Holt "Baby" 30 (Model 20-30 H.P.) and Holt 60, circa 1915.

Holt 75, from 1917 catalogue.

Illinois Army Reserve Holt 75 with Holt 20-ton wagon, circa World War I.

Canadian forces and Holt 75 in Palestine, circa World War I.

Holt 75 with Caterpillar trailers operating near the Mexican border, 1917.

Holt 75 in Salonika, Greece, 1915.

Holt 120, from 1917 catalogue.

Holt 120 bearing the Austrian double eagle insignia. Sold by Holt's Austrian dealer for purported agricultural use. However, it was most likely intended for military use, and was eventually used by the German army to haul heavy ordnance materials during World War I.

Camouflaged US Army Holt 120, circa 1917.

Holt 120 working in a cleanup operation following a flood in Peoria, Illinois.

Holt 18, 1915.

Holt 18 undergoing testing by US Army.

Holt 18 used by the Peoria US Army recruiting station, circa World War I.

Holt 12-18 (Holt 18).

Holt 45 artillery tractor, May 1917.

Holt 45 hauling a big gun into position, circa World War I.

Holt 45 Army tractor.

Armored Holt 45 Model E HVS with winch, circa World War I.

Holt Midget at the 1917 California State Fair.

Holt Midget at the 1917 California State Fair.

Holt Midget with a load of hay, circa 1919.

Holt Midget and Holt combine.

Holt 5 Ton with hand-controlled bulldozer.

Armored Holt 5 Ton.

A T-29 (5 Ton) moves uphill in front of dump trucks.

Holt T-29 and ripper on a road construction project.

Front view of an armored Holt 5 Ton.

Holt 5 Ton grading city streets in Dallas, Texas, 1922.

Holt T-11 (5 Ton) orchard tractor.

Holt 5 Ton. Note "USA" cast into the radiator.

Holt 5 Ton.

Holt 5 Ton equipped with Roadmaker Top, December 1924.

Holt 5 Ton hauling garbage, Dallas, Texas, 1924.

Holt 5 Ton threshing.

Holt 5 Ton drilling and seeding at the site of the Fargo, North Dakota State Fair.

Holt 5 Ton hauling an air compressor for the Pennsylvania Edison Company, Reading, Pennsylvania.

Holt 5 Ton switching railcars.

Holt 5 Ton in heavy rice land near Houston, Texas.

Holt 5 Ton plowing during a Wichita, Kansas demonstration.

Holt 5 Ton equipped with universal tracks and Roadmaker Top with winter enclosure, January 1925.

5 Ton pumping water for irrigation on a West Indies plantation, March 1929.

Holt 5 Ton and Maney scraper at work on Chicago Lake Front Park.

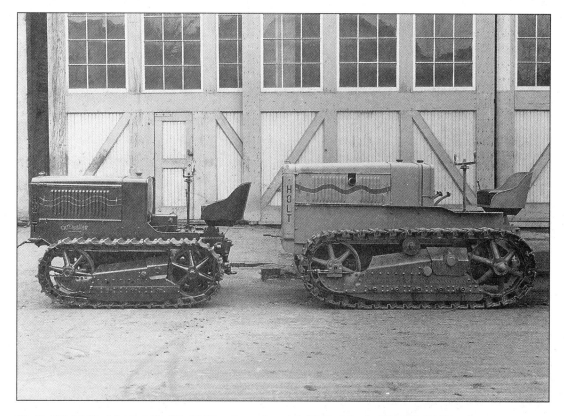

Holt 25 H.P. Model T-35 (2 Ton) and a 40 H.P. Model T-29 (5 Ton).

Holt "Western" 10 Ton.

Armored Holt 10 Ton undergoing testing by the US Army.

Armored Holt 10 Ton, Model 55 artillery tractor.

10 Ton and Russell elevating grader on a property development site, St. Louis, Missouri, 1919.

Two Holt 10 Ton tractors, each hauling three 3-1/2-yard Troy trailers, on an Erie, Pennsylvania job site, October 1919.

Holt 10 Ton with Holt-built land leveler, July 1920.

Holt 10 Ton hauling ashes in Philadelphia, Pennsylvania, 1921.

Holt "Northern Logger" 10 Ton on display in Milwaukee, Wisconsin during the September 1922 Northern Loggers Conference.

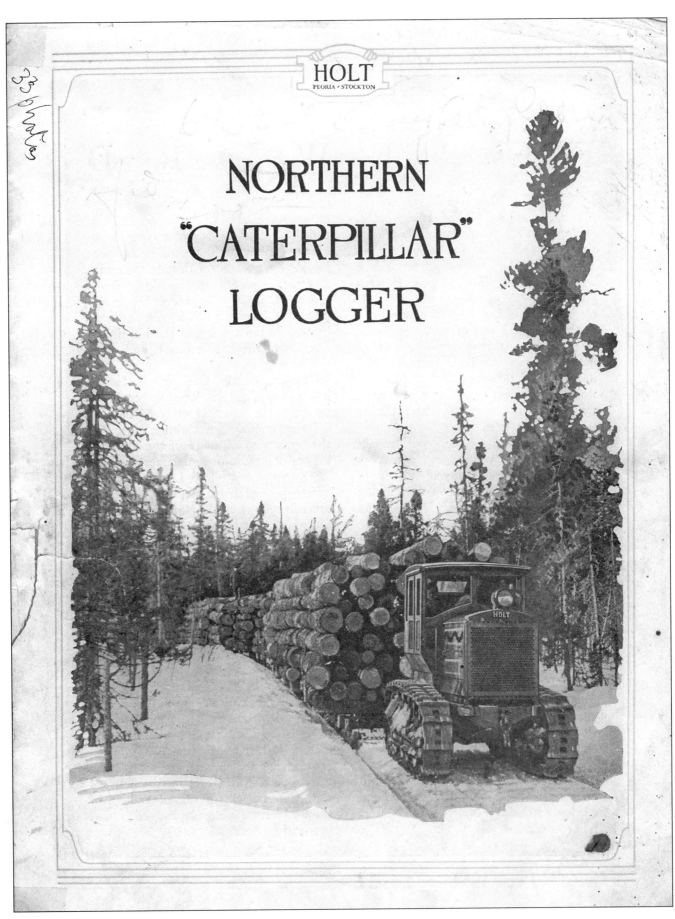

The cover of a sales brochure for the Holt "Northern Logger" 10 Ton.

Holt 10 Ton with grader during a demonstration at the University of Missouri, 1923.

Holt 10 Ton winching itself out of mud, while hauling oil field supplies, 1923.

Holt 10 Ton with Russell elevating grader working on Marquette Park, Chicago, September 1924.

Holt 10 Ton hauling six Troy trailers loaded with road building materials, Illinois, 1924.

Holt "Western" 10 Ton.

A Sinclair Oil Co. Holt 10 Ton with Troy trailers, hauling oil field supplies near Ranger, Texas.

A Sinclair Oil Co. Holt 10 Ton with Troy trailers, hauling oil field supplies near Ranger, Texas.

Holt 10 Ton hauling oil field equipment in Columbia, South America.

Holt 10 Ton working on removing 380,000 cubic yard hill for a Kansas City, Missouri water pumping station.

Holt "Western" 10 Ton and road grader.

Holt 10 Ton with winch lifting a 5,000 lb. block of concrete.

Front view of a Holt 10 Ton with winch.

Holt 10 Ton and Caterpillar trailer hauling sugar cane in Cuba.

Holt 10 Ton with 4500 feet of hardwood.

Holt 10 Ton and La Crosse disc plows operating on the King Ranch, Texas.

Holt 10 Ton with winch loading wagons in an Arkansas oil field.

Holt 10 Ton hauling supplies in Mayo, Yukon Territory, Canada.

Holt 10 Ton with Russell Mogul grader.

Holt 10 Tons, owned by Standard Oil Co., haul 35 tons of 6-inch pipe to the field.

Holt 10 Ton hauling sand on a property improvement project in Miami, Florida.

Holt 10 Ton skidding logs near Dakota City, Nebraska.

A lineup of 20 new Holt T-35 (2 Ton) tractors, 1921.

Side view of a Holt Model T-35 (2 Ton).

Front view of a T-35, June 1921.

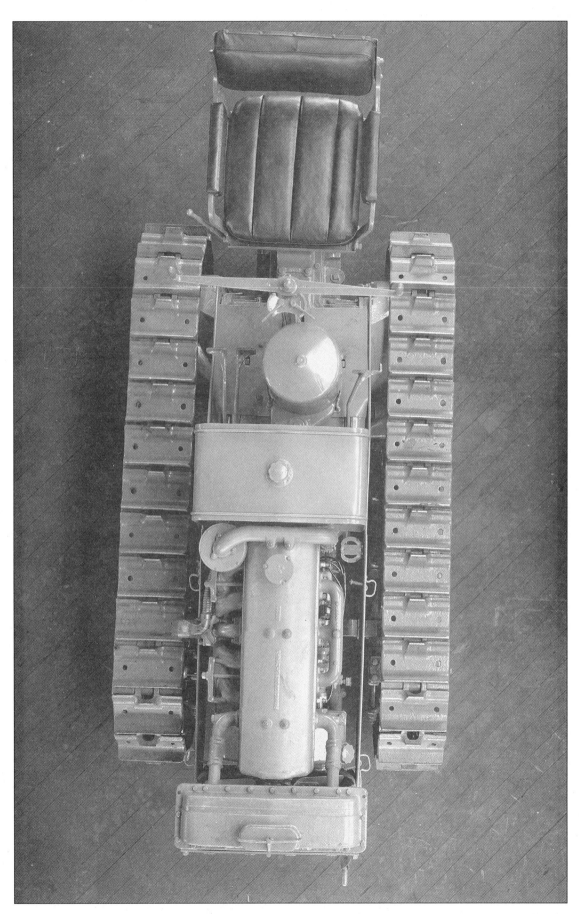

Overhead view of a Holt T-35 (2 Ton).

Holt T-35 (2 Ton) front view, June 1921.

Rear view of a Holt 2 Ton.

Holt T-35 pulling a small scraper.

T-35 pulling a wagon.

Holt 2 Ton yard locomotive.

Holt 2 Ton moving a shade tree in Wichita, Kansas.

Holt 2 Ton and Holt Model 39 harvester.

Holt 2 Ton and Fresno scraper, 1922. Note the team of five horses pulling the same unit.

Holt 2 Ton pulling rocks out of an orange grove, Ojai, California, February 1927.

Holt 2, 2.5, and 5 Ton tractors.

Cover from a catalogue of the C.L. Best Tractor Company, San Leandro, California.

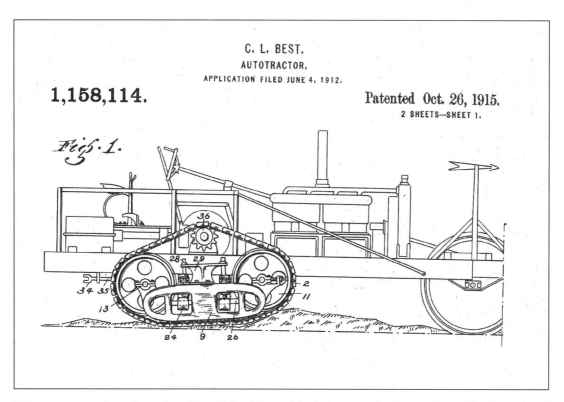

The patent drawing for the C.L. Best "Autotractor", from the 1912 patent application no. 1,158,114 (granted October 26, 1915).

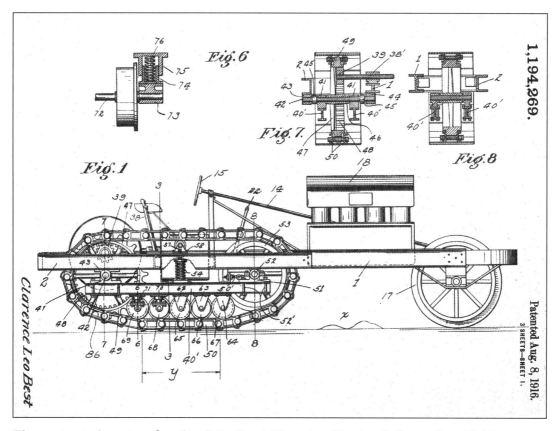

The patent drawing for the C.L. Best "Tractor Engine", from the 1913 patent application no. 1,194,269 (granted August 8, 1916).

Best 75.

Best 75 pulling a combine harvester near Walla Walla, Washington.

Best 75 hauling supplies to a lead ore mine in Death Valley, California, circa World War I.

Best 75 freighting across Nevada desert, circa World War I.

Best 75 photographed in 1954, then occasionally still used for plowing.

Best "Humpback" Thirty.

Drive gears on a Best "Humpback" Thirty.

Best "Humpback" Thirty with custom-built "coveralls" that permitted the tractor to work in the orchard without breaking limbs or knocking off fruit.

Best 40, from a 1916 catalogue.

Best 40, upon delivery to a Livermore Valley, California grain farmer.

Best 90, from a 1916 catalogue.

Best 90 hauling material in a tropical forest near Havana, Cuba.

Best 8-16 "Pony" Tracklayer, from a 1916 catalogue.

Best 25s (Model 16-25) in an asparagus field, Solano County, California.

Best Model 16-25 (Model 25) built in Dayton, Ohio during the time that C.L. Best Company was controlled by C. A. Hawkins of White Sewing Machines, White Trucks, and, later, Cleveland Tractor Company (Cletrac).

Rear view of a Best Model 16-25.

A Best 25 pulling a three-bottom plow.

Best Sixty hauling logs from a mill to the railroad, 1922.

Best Sixty being fueled at a gas station in San Leandro, California, July 1922.

Best Sixty with 6-bottom 14-inch plow, circa 1923.

Best Sixty in "difficult" terrain.

Best Sixty in "difficult" terrain.

Best Sixty with three-point Killefer subsoiler building a road into the Oakland, California Municipal Golf Course, August 1924.

Best Sixty freighting an 18-ton Dinkey engine a distance of 60 miles into the Utah mountains.

Best Sixty plowing snow with a Stroud grader.

Best Sixty and Adams Leaning Wheel Grader no. 12, with 12-foot blade and backsloper, Scott County, Iowa, 1925.

Best Sixty and line construction, 1925.

Best Sixty hauling gravel on a railroad maintenance project.

Best Sixty in a California logging operation.

Best Sixty during a grader demonstration.

Rear view of a Best Sixty with "California" top.

Best Sixty with elevating grader, on construction of UCLA Westwood, 1926.

Plowing snow with Best Thirty and Russell Standard grader, March 1923.

Best Thirty with LaPlant-Choate bulldozer, 1923.

Best Thirty moving a building, 1923.

Best Thirty plowing on a hillside.

Best Thirty.

A fleet of seven Best Thirty tractors on a farming operation, June 1924.

Best Thirty with disc harrow.

Best Thirty with Knapp plow in California cherry and prune orchard.

Best Thirty with Killefer front loader in use by the US Marines, 1926.

TRACKLAYER CREED

WE BELIEVE that a tractor should embody every quality that proves a profitable investment; should stand the most severe test of service, through a long period of time at minimum cost and maximum efficiency.

WE BELIEVE further, these are insured through the employment of high grade materials, by the most skilled workmanship and perfect design.

Finally costing more to build and more to buy, but less to own. Built in the "Best" shops on experience and progression—backed by the "Best" reputation.

From a catalogue of the C.L. Best Tractor Company, San Leandro, California.

The cover of a 1925 sales brochure for the Caterpillar 10 Ton.

Caterpillar 5 Ton pulling a combine harvesting barley on the farm of C. Parker Holt, vice-president of Caterpillar Tractor Co., July 1929.

Caterpillar 5 Ton, circa 1926.

Rear view of a Caterpillar 2 Ton, showing belt pulley, November 1926.

Caterpillar 2 Ton, 1927.

Caterpillar 2 Ton.

Caterpillar 2 Ton, 1927.

Caterpillar 2 Ton on a factory test bed, 1927.

Caterpillar 2 Ton and dump wagon, 1927.

Russell Motor Patrol No. 4, powered by Caterpillar 2 Ton, with standard 12-foot blade and optional snow plow attachment, February 1927.

Caterpillar 2 Ton pulling stumps, Polk County, Wisconsin, June 1927.

Caterpillar 2 Ton moving a bolder on the grounds of Essex County Country Club, West Orange, New Jersey, July 1927.

Caterpillar 2 Ton at the Peoria Service Training School, 1928.

Caterpillar 2 Ton handling culvert pipe.

Caterpillar 2 Ton with rubber track used for general hauling at a factory in Hull, England, 1928.

Caterpillar 2 Ton with rubber track used for general hauling at a factory in Hull, England, 1928.

Caterpillar 2 Ton pulling stumps out of an old vineyard, March 1928.

Caterpillar 2 Ton and the Capetown, South Africa dealer, 1928.

Caterpillar 2 Ton, fitted with ridge breakers, working a California grove, 1928.

Caterpillar 2 Ton with leveling drag working on the aerodrome site in Hull, England, 1929.

Caterpillar 2 Ton supplying the power for a sawing operation, Albermarle County, Virginia, June 1930.

Caterpillar Sixty with an asparagus chopper, California, January 1927.

A custom tractor based on a Caterpillar Sixty, and used to prepare ground for drying salt, Alviso, California, September 1927.

Caterpillar Sixty with Rumley combine, Colusa, California, October 1927.

Caterpillar Sixty Logging Cruiser, June 1928.

Caterpillar Sixty Logging Cruiser, June 1928.

Caterpillar Sixty with LaPlant-Choate bulldozer, Graham, North Carolina, 1928.

Caterpillar Sixty with no. 68 Killefer revolving scraper, August 1928.

Caterpillar Sixty pulling a 7-yard crawler wagon, August 1928.

Caterpillar Sixtys with track arches operating near Westwood, California, October 1928.

Caterpillar Sixty with Hauser brush cutter.

Caterpillar Sixty hauling ties from a lumber operation in Hudson, Ontario, Canada, April 1929.

Caterpillar Sixty with Gardner loader, Long Beach, California, May 1929.

Caterpillar Sixty providing power to a stationary thresher, Onawa, Iowa, September 1929.

Caterpillar Sixty with LaPlant-Choate bulldozer building drainage ditches for the Pennsylvania Railroad, Metuchen, New Jersey, December 1929.

Caterpillar Sixty fitted with factory cab, lights, and Willamette winch, 1930.

Caterpillar Sixty ripping up ground ahead of a grader, Mt. Hood National Forest, Oregon, 1930.

Caterpillar Sixty logging with Fairlead wheels, Kinzua, Oregon, 1930.

Caterpillar Sixty with Master backfiller operated by a Willamette double drum winch, September 1930.

Wide-track Sixty pulls a disk harrow in land reclaimed from Holland's Zuider Zee, 1931.

Caterpillar Sixty with LaPlant-Choate snowplow, Itasca, Illinois, March 1931.

Caterpillar Sixty logging with a pan near Gilmer, Washington, May 1931.

Caterpillar Sixty cleaning lake front at Oak Street beach, Chicago, Illinois, May 1931.

Caterpillar Sixty and 8-bottom John Deere plow in the Palouse Hills near Pullman, Washington, October 1931.

Caterpillar Sixty bulldozing garbage, Newark, New Jersey, October 1931.

Caterpillar Sixty moving a house, circa 1932.

Caterpillar Thirty with LaPlant-Choate hydraulic bulldozer, July 1927.

Three Caterpillar Thirtys remove debris left by flood along the Santa Clara River, following the St. Francis Dam collapse, March 1928.

Three Caterpillar Thirtys used in removing silt washed onto land by the flood that followed the St. Francis Dam collapse of March 1928. Hundreds died and more than 29,000 acres of orchards and crops were damaged.

Caterpillar Thirty with a root plow, the second operation in preparing the subgrade on the Peoria Trail, Peoria, Illinois, August 1928.

Caterpillar Thirty tows a railcar toward a B&O car repair shop, Garrett, Indiana.

Caterpillar Thirty with high wheeler logging on the Klamath Indian Reservation, Oregon, October 1928.

Caterpillar Thirty with Holt combine, Faulkton, South Dakota, October 1928.

Caterpillar Sidehill Thirty with Killefer double disc, Vacaville, California, February 1929.

Rear Seat Hill Special Caterpillar Thirty in a Watsonville, California orchard, April 1929.

Caterpillar Thirty with Willamette double drum winch rigging a Texas oil well, April 1929.

Caterpillar Thirty equipped with a Highway post hole digger, December 1929.

Caterpillar Thirty equipped for winter work, December 1929.

Caterpillar Thirty with rear-mounted seat.

Caterpillar Thirty pulling 20 cords of pulp wood, Port Arthur, Ontario, Canada.

Caterpillar Thirty with optional bumper and canopy.

Thirty with Killefer loader operating in a Chicago, Illinois brickworks.

Caterpillar Thirty with three Euclid scrapers on a state highway project, Loudoun County, Virginia, April 1930.

Caterpillar Thirty equipped with Highway trailer reel cart and winch, Ashland, Virginia, June 1930.

Caterpillar Thirty with Euclid wagon at work on the new stadium at the University of Virginia, Charlottesville, October 1930.

Caterpillar Thirty and a Ball wagon grader building a road near Seville, Georgia, June 1931.

Caterpillar Thirty and rotary scraper working on a highway through Bad Lands National Park, South Dakota, August 1931.

Caterpillar Thirty with a Hi-Way Service scarifier, ripping up old road prior to resurfacing, August 1931.

Wide Gauge Caterpillar Thirty pulling four Athey wagons loaded with sugar cane, Lake Okeechobee, Florida, February 1933.

Caterpillar Twenty and disc near Hastings Ranch, California, September 1928.

Caterpillar Twenty pulling a potato digger and Tusco bagger near Farmingdale, New Jersey, October 1929.

Caterpillar Twenty leveling irrigation checks near Oakland Airport, July 1930.

A Caterpillar Twenty brings in the hay.

Caterpillar Twenty with cable plow.

Caterpillar Twenty with automatic spring-tooth cultivator at work in an orange grove near Placentia, California, June 1931.

Caterpillar Twenty with disc breaking up guano during the annual guano harvest on Island Don Martin off the coast of Huacho, Peru, 1931.

Caterpillar Twenty pulling sacks of guano during the annual harvest on Island Don Martin off the coast of Huacho, Peru, 1931.

Caterpillar Twenty with Allsteel sideboom handling rail on a rail relaying job for the Chicago & Great Western Railroad, Peal City, Illinois, 1932.

Caterpillar Twenty with Allsteel sideboom tearing up abandoned track and loading old ties on tie car for the Nickel Plate Railroad, East Peoria, Illinois.

Caterpillar Twenty clearing out Farm Creek, East Peoria, Illinois, in a work relief program of February 1932.

Caterpillar Twenty with double drum Willamette winch pulling stumps along Farm Creek, East Peoria, Illinois, March 1932.

Two Caterpillar Twentys and a Caterpillar Thirty with artillery.

Caterpillar Twenty spreading coal at the Saginaw River, Michigan plant of Consumer's Power Company, June 1932.

Caterpillar Twenty spreading coal at the Saginaw River, Michigan plant of Consumer's Power Company, June 1932.

The cover from *Abraham Lincoln on the coming of the Caterpillar Tractor*, a 1929 publication reproducing a Lincoln speech of September 1859.

New Model "Caterpillar" Ten Tractor

SEVENTY YEARS AGO

ABRAHAM LINCOLN SAID:

"Our Thanks, and Something More Substantial than Thanks, are due to Every Man engaged in the Effort to Produce a Successful Steam Plow."

DEDICATION

—

The prophecy and hope of Abraham Lincoln, made in 1859, [see page 15] was fulfilled by pioneer spirits such as Daniel Best, Benjamin Holt, and their respective sons.

The result of their genius and invention is again viewed in the latest model of the successor to the "Steam-Plow"—the New "Caterpillar" Ten Tractor—which today takes its place in Agriculture and Industry with the "Caterpillar" Models 2-Ton, Twenty, Thirty and Sixty.

Joseph A Quinn

Caterpillar Ten orchard tractor disking in an orange grove, Azusa, California, April 1929.

Caterpillar Ten orchard tractor disking in an orange grove, Azusa, California, April 1929.

Caterpillar Ten operating a 10-inch centrifugal pump with 9-foot lift, irrigating sugar beets, June 1929.

Caterpillar Ten and Olson disc cutting brush in a grape vineyard near Fresno, California.

Caterpillar Ten road maintainer.

Caterpillar Ten pulling a New Idea transplanter in cabbage, Kent, England.

Fueling a Caterpillar Ten, 1930.

Caterpillar Ten with Caterpillar Russell grader on a Minnesota roadway, 1930.

Caterpillar Ten with 2-row New Idea corn picker, October 1930.

Caterpillar Ten hauling sleds of lettuce, Waco County, Oregon, June 1930.

Caterpillar "High" Ten with experimental front-mounted spring-tooth cultivator, 1932.

Thirteen Caterpillar's at work for a movie being shot on the 3900-acre Paschendal Dairy Farm, north of Chicago. The farm operated four High Clearance Tens, six Fifteens, two Twentys, and a standard Ten, 1932.

Wide gauge 1929 Caterpillar Ten (photographed in 1969).

Caterpillar Fifteen orchard tractor in California orchard, April 1928.

Caterpillar Fifteen sulphuring a grape vineyard near Clovis, California, 1930.

Caterpillar Fifteen with screen hood over engine, July 1930.

Caterpillar Fifteen pulling a low-built toboggan-type sled in a Hood River Valley, Oregon apple orchard, October 1930.

Caterpillar Fifteen and road grader on a Minnesota road.

Russell Fifteen Motor Patrol powered by Caterpillar Fifteen tractor.

Caterpillar Fifteen hauling coke and supplies 26 miles over trails to a gold mining camp. The tractor crossed two mountain ranges and operated on grades up to 17%. Azurite Gold Co., Mazama, Washington, October 1931.

Two Caterpillar Fifteens freighting supplies to a gold mining camp. Azurite Gold Co., Mazama, Washington, October 1931.

Caterpillar Fifteen providing power to a mill on an Illinois farm, December 1932.

Caterpillar Fifteen engine.

Caterpillar Fifteen with 1/2-yard bucket "Shovel Loader" built by Anderson Brothers, Port Richmond, Staten Island, New York, December 1932.

Caterpillar Fifteen plowing under a cover crop, Hungerford, England.

Caterpillar Fifteen and Caterpillar 34 combine.

The first Caterpillar diesel engine in a Caterpillar Sixty, 1930.

Caterpillar Diesel Sixty (serial no. 1-C-2) ripping earth with a Brenneis chisel, Woodland, California, April 1932.

Caterpillar Diesel Sixty (serial no. 1-C-7) and blade grader, May 1937.

Caterpillar Diesel Sixty (1-C-5) with LaPlant-Choate and Trackson crawler wagons on Mississippi River levee construction, Talullah, Louisiana.

Diesel Sixty and 8-yard 3-way dump wagon on a coal stripping operation, Shenandoah, Pennsylvania, June 1932.

Diesel Sixty with Hyster scraper digging a wider and deeper channel in Mill Creek, Walla Walla, Washington, June 1932.

Caterpillar Diesel Sixty on construction of the Bouquet Canyon Dam, California, December 1932.

Three Diesel Sixty-Fives and a gas-powered Sixty (center/rear) pulling sheepfoot tampers on construction of the Bouquet Canyon Dam, 75 miles north of Los Angeles, California, December 1932.

Caterpillar Sixty-Five.

Caterpillar Twenty-Two pulling a PTO-driven rice binder near Butte City, California, October 1934.

Caterpillar Twenty-Two hauling logs from a New Hampshire logging operation, July 1935.

Caterpillar Twenty-Two and 3-point Killefer ripper in a Napa Valley orchard, May 1934.

Caterpillar Twenty-Two with 6-1/2-foot Killefer disc operating in an orchard with grades from 60 to 88%, Vacaville, California, May 1934.

Caterpillar Twenty-Two cultivating corn near Holyoke, Colorado, August 1934.

Caterpillar Twenty-Two pulling a homemade leveler, San Luis Rey, California, January 1938.

Caterpillar Twenty-Five with wide tracks working on site of an airfield at the Quantico Marine Barracks, Virginia. A Caterpillar Fifteen with rotary scraper and a Caterpillar Thirty with bulldozer are also at work, November 1932.

Hartburg Sugar Beet Piler built around a Caterpillar Twenty-Five. The Twenty-Five operated the piler's mechanism and propelled the machine on non-Caterpillar tracks. Great Western Sugar Co., Denver, Colorado, October 1932.

Caterpillar Twenty-Five with Hughes-Keenan Roustabout Crane working in the Union Pacific yards, Cheyenne, Wyoming, March 1932.

Caterpillar Twenty-Five with Hughes-Keenan Roustabout Crane towing sets of flat car wheels, each set weighing 2400 lbs. Union Pacific yards, Cheyenne, Wyoming, March 1932.

Caterpillar Fifty with Master Trailbuilder building a road for the US Forest Service, 10 miles southwest of Randle, Washington. June 1932.

Caterpillar Fifty and 8-yard wagon hauling from a Bucyrus-Erie 1-1/4-yard Gas+Air shovel on a state highway construction project near East Columbus, Ohio, May 1932.

Caterpillar Fifty with Cardwell-Allsteel oilfield winch on a Kansas well, July 1932.

Caterpillar Fifty and Athey wagon hauling from a Lorain 75B shovel on road construction near Columbus, Ohio, September 1932.

Caterpillar Fifty with Isaacson Roadbuider at work on the Mt. Baker Park Highway in Washington, September 1932.

Diesel Fifty being fueled, 1934.

Diesel Fifty and elevating grader building roads through a cornfield, 1934.

Diesel Fifty pumping a Rainmaker that supplied irrigation to 500 acres of early pink beans, Robbins, California, August 1934.

Caterpillar Thirty-Five with Blaw-Knox bulldozer and ATECO sheepfoot tamper on a road project west of Waukegan, Illinois, June 1932.

Caterpillar Thirty-Five and 30-inch Hester Fire Break plow building a fire break near Savannah, Georgia, September 1932.

Caterpillar Thirty-Five with Allsteel rod and tubing winch at an oil derrick near Wichita, Kansas, November 1932.

Gasoline Thirty-Five (R35) with brushpuller and rake designed by the US Forest Service mounted on a Isaacson frame. The tractor was equipped with 24-inch shoes for working on soft ground in the Kanikusa National Forest near Priest River, Idaho, August 1934.

Caterpillar Thirty-Five and Caterpillar grader cutting ditches on country roads in northern Alabama.

Caterpillar Thirty-Five and chisel plow working in California hills.

Caterpillar Thirty-Five and Caterpillar grader operating in France.

A renovated Caterpillar Thirty-Five clearing land in the late 1940s.

Diesel Thirty-Five with two 10-foot Emerson disc weeders summer fallowing wheat land near Hudson, Colorado, August 1933.

Diesel Thirty-Five pulling a Caterpillar no. 40 grader, cleaning and making irrigation ditches near Phoenix, Arizona, April 1934.

Diesel Thirty-Five and 6-row Shurmule ridge buster on the Hays Experimental Farm, Hays, Kansas, May 1934.

Diesel Thirty-Five operating an irrigation pump with 190-foot lift, July 1934.

Diesel Thirty-Five with Master bulldozer building a canal for running water used in a gold dredging operation near Sacramento, California, July 1934.

Diesel Thirty-Five opening gates at an irrigation site.

Diesel Thirty-Five hauling Athey track wagons filled with sugar cane, Florida, February 1935.

Diesel Thirty-Five combines barley in the San Joaquin Valley, California, 1945.

Caterpillar Twenty-Eight and John Deere swamp plow on Saghalien Isle, Japan, September 1935.

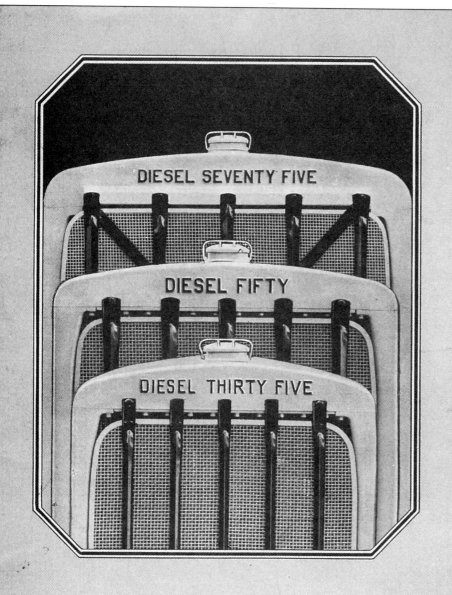

Diesel Seventy-Five and a Fowler four-bottom reversible plow in northern France, August 1933.

Diesel Seventy-Five with LeTourneau rubber-tired "buggy" on a San Francisco-Oakland Bay Bridge approach, 1934.

Diesel Seventy-Five pulling a 5-disk heavy plow with 32-inch discs and a 12-foot Killefer cover crop disk harrow, October 1935.

Diesel Seventy-Five pulling a heavy Waco disc plow.

Diesel Seventy-Five with Hyster arch logging out redwood, California, 1938.

Diesel Seventy-Five with 8-section sheepfoot tamper on dam project, 1940.

Caterpillar Seventy-Five clearing land with 300 feet of 1-1/2-inch chain hooked to another Seventy-Five (not shown).

Diesel Forty grading a road.

Diesel Forty working around the San Francisco-Oakland Bay Bridge, 1937.

Caterpillar Thirty (6G Series) discing in a California pear orchard, April 1936.

Caterpillar Thirty (6G Series) skidding logs near Lisbon, Ohio, May 1936.

Caterpillar Thirty (6G Series), 1937.

Caterpillar Thirty Orchard Tractor (6G Series), 1937.

Butane Caterpillar Thirty (6G Series).

Operator controls of a Caterpillar Thirty (6G Series).

Caterpillar Thirty (6G Series) with Anthony Model B loader, June 1937.

Caterpillar Thirty (6G Series) with Trackson sideboom and winch moving tank sections for a new refinery, Lovell, Wyoming, August 1937.

Caterpillar Thirty (6G Series) with Trackson loader excavating a basement, East Peoria, Illinois, November 1937.

Caterpillar Thirty (6G Series) and Rumley combine, Stafford County, Kansas, July 1937.

Caterpillar Thirty (6G Series) and Killefer rotary scraper, January 1938.

Caterpillar R2 switching railcars.

The cover from a D2 sales brochure, circa 1940.

D2 and 3-row furrower in a fig grove near Reedly, California, June 1938.

D2 and Killefer 6-foot offset cover-crop disc, in an avocado orchard, Vista, California, April 1939.

D2 with a 300-gallon Hardie spray rig in tomatoes. Berrien Springs, Michigan, June 1939.

D2 with Atlas Tumblebug constructing new terraces for an orange grove. East Highlands, California, January 1948.

The cover from a sales brochure for the D2 and D4, circa 1945.

D2 with Killefer pan breaker subsoiling 365 acres of hop land. Yuba City, California, October 1948.

D2 with Killefer chisel renovating Bermuda grass in a Valencia orange grove, Scottsdale, Arizona, November 1948.

D2 pulling a hay rack for feeding cattle following a blizzard. Oshkosh, Nebraska, February 1949.

D2 with Rome disc harrow in an orchard near Bedford, Virginia, March 1949.

D2 and John Deere 8-foot disc in a cover crop of rye in an orchard in Hart, Michigan, May 1949.

D2 with liquid fertilizer spreader in an almond orchard near Yuba City, California, November 1949.

D2 with Trackson loader on a California housing project, 1949.

D2 Citrus tractor and offset disc in a 3-year old navel orange grove, Arlington, California, 1949.

D2 with no. 41 John Deere moldboard plow near Phelps, New York, April 1950.

D2 with Trackson LW2 TracLoader loading fertilizer into a spreader. Poughkeepsie, New York, April 1950.

D2 with Graham Hoeme tiller seeding and fertilizing in irrigated wheat land. Moses Lake, Washington, March 1952.

A Caterpillar dealer display of "Bonded Buy" program D2s. This quality-assurance program for used Caterpillar equipment dated from the early 1950s.

D2 with Minneapolis-Moline plows cultivating a vineyard of Italian Swiss Colony, Asti, California, April 1955.

On its way to the South Pole. A Low Ground Pressure D2 bundled for a parachute drop from an 18th Air Force C-124 Globemaster.

Low Ground Pressure D2 dropped from a C-124 Globemaster to a camp at the South Pole.

Low Ground Pressure D2, following South Pole parachute drop, hauls load of fuel barrels from drop area to camp.

Sideboom D2 tractor on a utility job site. All controls were hydraulically operated from the pump on the front of the tractor.

Sideboom D2 tractor featured a 1200 lb. counterweight. It had 8400 lb. lift capacity at 2-foot overhang.

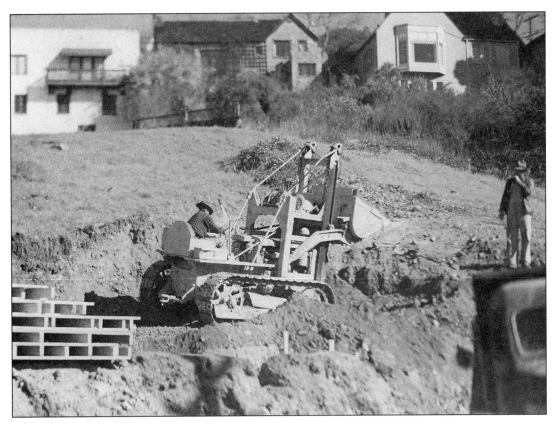

D2 with front lift-and-dump shovel excavating basement foundations and sidewalks on a home building job, East Bay, California, circa 1959.

Caterpillar tool bar for use on D2 (as shown) or D4.

The cover from a sales brochure for the D4, circa 1940.

RD4 pulling a homemade float, preparing 4000 acres of land for tomatoes, Brentwood, California, April 1937.

RD4 with LeTourneau Angledozer spreading sawdust to be burned by a Montana power plant, October 1937.

RD4 combining wheat with a John Deere 35 combine, Endicott, Washington, August 1938.

RD4 with front-mounted ripper making a road bed through a Hawaiian sugar plantation.

RD4 in an Imperial Valley, California cabbage field, circa 1940.

A D4 with charcoal burner pulled a disc plow in Tunisia, circa World War II.

D4 with LeTourneau sheepfoot tamper and no. 12 Motor Grader building an airstrip in New Guinea, February 1943.

D4 on Attu Island in the Aleutians, May 1943.

Hooking a 37 mm anti-tank gun to a D4, Massacre Bay, Attu, Aleutian Islands, May 1943.

D4, November 1944.

D4 pulling a platform from which workers pruned fronds of Deglet Noor date palms. Indio, California, April 1946.

D4 orchard tractor pulling a Killefer offset disc in an almond orchard near Durham, California, April 1947.

D4 clearing rock in an apple orchard, Chazy, New York, September 1947.

D4 with no. 4S hydraulic bulldozer pulling a 1-shank subsoiler as it prepares old tomato land for avocados. Vista, California, January 1948.

D4, with specially built shaker, shakes up to eight acres of walnuts per day, San Jose, California, October 1948.

D4 with 4-ton Killefer subsoiler preparing land for a new orchard for the Veteran's Home of California, Napa, California, October 1948.

D4 with Trackson loader on a building site.

D4 with no. 4S bulldozer cutting down a bank for improvement of highway frontage, San Jose, California, April 1949.

D4 with no. 4A bulldozer and no. 40 scraper in a soil erosion dam construction project, July 1949.

D4 with tool bar equipped with four lister bottoms at the Peoria Proving Grounds, September 1949.

D4 with no. 4S bulldozer at work on a stock tank, Amarillo, Texas, 1949.

R4 with LeTourneau dozer, circa 1950.

D4 with Trackson Traxcavator loading sand for Santa Cruz County, Arizona, April 1950.

D4 with saddle bags in winter sugar beet harvest, Sacramento Valley, California, January 1951.

"Want to trade?" D4 and an early Holt crawler at a California dealership.

A dairy farmer uses his D4 to plow a path through snow to his stranded cows. Wokley, Idaho, February 1954.

D4 skidding logs in a forest southeast of Hamburg, Arkansas, October 1956.

Low Ground Pressure D4 pulling 10 sleds heads for its Antarctic base camp.

At Hut Point, McMurdo Sound Air Base, this Hystaway-equipped D4 sets up part of a ground control approach unit to guide in landing aircraft.

Caterpillar D4 with orchard seat and controls.

RD6 and John Deere combine harvesting wheat near Reardon, Washington, August 1938.

The cover from a sales brochure for the D6, circa 1940.

RD6 with John Deere Van Brundt seeders planting wheat in western Idaho.

Two RD6s and a Diesel Fifty combine rice in Imperial Valley, California, November 1941.

Caterpillar DW10 tractor with LaPlant-Choate CW10 scraper, and D6 with LeTourneau dozer, from the June 1941 Caterpillar Road Show.

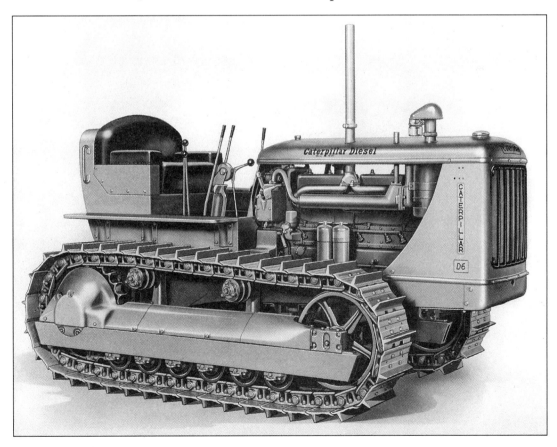

D6, November 1944.

D6 with no. 606 John Deere plow in a lettuce field near El Centro, California, June 1947.

D6 with Caterpillar Cable Control pulling a Caterpillar no. 60 scraper, 1947.

D6 with Trackson Traxcavator loading tungsten ore in a Nevada mining operation, April 1948.

D6 with Trackson Traxcavator loading tungsten ore in a Nevada mining operation, April 1948.

D6 with Trackson Traxcavator loading tungsten ore in a Nevada mining operation, April 1948.

Portable drill on a D6 with Winterweiss compressor.

D6 wheel tractor converted by a Caterpillar dealer and used for pulling circus wagons for Ringling Brothers Barnum and Bailey Circus, Saratoga, Florida, May 1951.

D6 with Balderson inside-mounted blade doing site preparation work.

D6 with three 8-foot Noble cultivators and a 24-foot drag harrow doing summer fallow work on an 1800-acre Montana wheat ranch, June 1956.

D6 with a Hyster winch skidding logs near Pagosa Springs, Colorado, 1956.

The cover from a sales brochure for the D7, circa 1950.

RD7 with four 16-inch plows follows a RD4 with two 16-inch bottoms in bean land in Camarillo, California, November 1936.

RD7 pulling a Marvin land plane, Knights Landing, California, October 1938.

RD7 with vee-dozer.

RD7 with vee-dozer.

RD7 with LeTourneau scraper on roadwork in Arizona, where the last unpaved link of roadway between East and West was being closed, 1940.

D7 with LeTourneau bulldozer building a levee along the Alamo River near Niland California, December 1940.

D7 with LaPlant-Choate bulldozer clearing London streets of debris following an air raid, August 1942.

D7 and grader at work on Rendova Island, Solomons, 1943.

An armored D7 with LaPlant-Choate Trailbuilder blade widening a road in France, circa 1944.

A US 7th Army armored D7 with LeTourneau Angledozer clearing German vehicles destroyed by troops of the US 3rd Division in their drive for Montelimar, France, December 1944.

D7, November 1944.

D7 clearing mesquite with LaPlant-Choate bulldozer frame to which was mounted a special Winchtex blade for uprooting and pushing over trees. Stamford, Texas, February 1946.

Two D7s, each pulling one end of a 300-foot U-shaped loop of heavily weighted 1-1/2-inch cable, in a land clearing project in the Dominican Republic, 1946.

D7 with Caterpillar Cable Control pulling a Caterpillar no. 70 scraper through an eroded gully being filled prior to cutting a new drainage channel, June 1947.

D7 with front-mounted stump rooter built by Florida Land Clearing Equipment Co., circa 1948.

D7 with Traxcavator loader.

D7 with Rome KG blade clearing trees.

D7 and Caterpillar no. 70 scraper leveling an emergency airstrip for the US Forest Service, in the Cascade Mountains near Coles Corner, Washington, June 1949.

Two D7s with Hyster winches skid logs to a landing, while a third D7 with no. 7A bulldozer piles logs near Ukiah, Oregon, September 1950.

D7 with no. 7S bulldozer wading into surf to extend line from winch to anchor a causeway, during practice for amphibian landing, September 1950.

A Caterpillar Diesel no. 12 Motor Grader and D7 extending the runway at Ebee Field, Fort Belvoir, Virginia, October 1950.

D7 pulling a John Deere combine, while a D6 pulls a bankout wagon in a rice harvesting operation near Butte City, California, October 1952.

D7 with no. 7S bulldozer and Hyster winch skidding logs from woods to landing, on a logging operation near Myrtle Point, Oregon, September 1956.

D7 with a no. 7S bulldozer on a levee construction project.

D7 with Gyro Dozer, the four penetrating teeth of which allowed the blade a combined function of ripping and moving material in one bulldozing operation. This unit was displayed at the 1957 American Road Builders Association show.

John Wayne at the controls of a D7 used in making the movie "Hellfighters."

The cover from a sales brochure for the D8, circa 1947.

RD8 with Isaacson bulldozer.

D8 clearing land near Villa, Texas, April 1938.

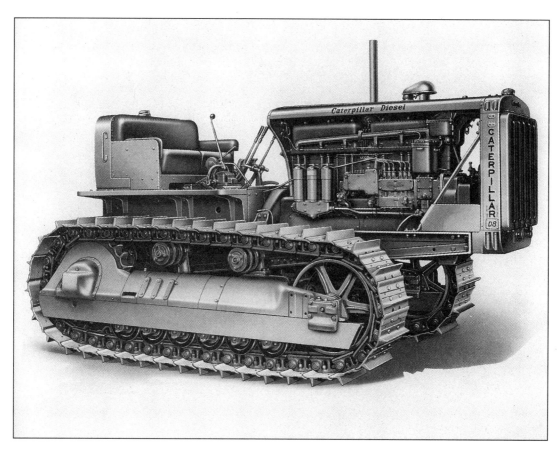

D8, circa 1941.

D8 and LeTourneau Carryall scraper cutting down a slope and making fill for Golden Gate National Cemetery, January 1941.

D8 and LeTourneau Carryall Scraper, one of a fleet moving sandy loam to fill a canyon on the site of San Francisco State College, Lake Merced, California, January 1941.

D8 bulldozing the fire of a burning B-29 to push dirt over flames, following a Japanese attack of Saipan, circa World War II.

A Caterpillar dealership in 1946. Shown are a D2, two D4s, and a D8.

D8 with LeTourneau CK8 Angledozer stripping overburden in Dutch Guinea, October 1943.

D8 with LaPlant-Choate Treedozer clearing land near Ben Hawkins, Texas, February 1946.

D8 with a 5-disc plow operating in a sand/clay mixture near Brownfred, Texas, May 1948.

D8 with air compressor and drill used in pioneering work in an open pit copper mining operation, Ray, Arizona, 1949.

D8 with no. 8A bulldozer building a road for a copper ore mining operation, Ray, Arizona, March 1949.

D8 with no. 8A bulldozer building roads in California. March 1949.

D8 with an 8-foot root plow subsoiling hardpan at an 18-inch depth. San Beneto, Texas, August 1949.

D8 with Caterpillar no. 80 scraper cutting hills and filling valleys for a cemetery in Whittier, California, November 1949.

D8 with Hyster logging arch and winch pulling logs to the landing in a Flea Valley logging operation, Stirling City, California, August 1950.

D8 with no. 8S bulldozer and D7 with no. 7A bulldozer in an open pit molybdenum mine, Climax, Colorado, August 1950.

D8 with no. 8S bulldozer in an open pit molybdenum mine, Climax, Colorado, August 1950.

D8 and no. 8S bulldozer at an open pit molybdenum mine, Climax, Colorado, August 1950.

D8 and no. 8S bulldozer at an open pit molybdenum mine, Climax, Colorado, August 1950.

D8 with cables holding pipe in place.

Two D8s pull, as one D8 pushes, on an unidentified construction project.

D8 backfilling a trench.

D8 on a logging road culvert construction project.

D8 bulldozing a railroad right-of-way.

D8 and Hyster Hystaway with grapple hook stacking hay in grazing land northwest of Winnemucca, Nevada, July 1951.

D8 with bank sloper attachment, developed to shore banks in irrigation canal building. Bakersfield, California, 1951.

Caterpillar DW21 tractor with no. 21 scraper and D8 pusher on a highway construction job between Portland, Oregon and Seattle, Washington, 1953.

D8 on an Oregon highway project, October 1953.

Two DW20 tractors with no. 20 scrapers and D8s working on a 6-lane approach to the Richmond San Raphael Bridge, Richmond, California, March 1955.

Caterpillar Diesel no. 583 Pipelayer, a D8 with Pipelayer Sideboom, and D7 with Pipelayer Sideboom working on pipeline construction. Crossing the Lackawaxen River at Hawley, Pennsylvania, July 1955.

D8 with no. 80 scraper push loaded by a D8 with push cup, on section of the Missouri Freeway near Joplin, Missouri, June 1956.

D8 with no. 8A bulldozer and Hyster winch skidding logs near Foresthill, California, August 1956.

D8 pulling five 7-foot Noble seed drills, planting winter wheat near Carter, Washington, September 1956.

D8 pulling five 7-foot Noble seed drills, planting winter wheat near Carter, Washington, September 1956.

D4s and D8s, built from US and UK parts and subassemblies, at a Caterpillar dealership in England, August 1956.

A British-built D8 and Harmon S. Eberhard, president of Caterpillar when this photo was taken in 1956.

D8 pulling a Euclid truck out of a hole. A Northwest hoe waits nearby, 1956.

Refueling Low Ground Pressure D8 of the US Army Engineering Arctic Task Force. January 1957.

Low Ground Pressure D8 working on an ice runway at McMurdo Sound.

Low Ground Pressure D8, equipped with 54-inch track shoes and cold weather protected to minus 65°F, tows a US Air Force C-47 that crash-landed near Greenland Ice Cap.

D8 with no. 8S bulldozer and Fleco root plow in pasture land near Wichita Falls, Texas, March 1958. Note the seeder box at left of the operator's seat.

D8 with rock plow in coral lime rock where tomatoes were to be planted. Three cuts were required to reach the desired depth. Goulds, Florida.

D8 with no. 463 scraper, 1958.

D8 towing a steel ball used to roll down trees and brush growth on the site of Hungry Horse Dam, Montana.

"Siamese" D8s clear land for the Hungry Horse Dam project, Montana.

D8s pulling LeTourneau and Caterpillar scrapers on the Hungry Horse Dam project, Montana.

D8 Siamese "Twin D8" and friend, a toy D4 peddle tractor.

Underside of a "Twin D8" with Holt Funnel Dozer.

First phase of an Arctic airlift in support of the DEW line "Operation Big Haul". D8s, Caterpillar 6 Traxcavators, and Caterpillar no. 12 Motor Graders.

D8 unloaded from a C-124 Globemaster of the 18th Air Force Tactical Air Command, part of "Operation Big Haul".

D8H with ATECO cable plow laying cable, May 1963.

Wooden mockup of a D9 used over the course of the D9's 10-year development period. Photo from the Caterpillar Research Department.

Experimental D9 (D9X) and D8 being prepared for field testing, 1954.

Experimental D9 (D9X) aboard a railcar on its way to the field for testing, 1954.

Experimental D9 (D9X) self loads a no. 90 scraper, 1954.

D9 plowing in Sutter Basin, California.

D9 with 14-foot root plow and seeder clearing range land in Texas.

D9 with no. 9A blade clearing 3 acres of range land/hour on the King Ranch, Kingsville, Texas, December 1955.

DW21 tractor with no. 470 scraper push loaded by a D9 with no. 9S bulldozer, preparing roadbed for Great Eastern & Pacific Railroad, British Columbia, Canada, August 1956.

9U bulldozer for D9 tractor, January 1957.

D9 with no. 9 ripper, 1958.

D9 with power shift at work. Power shift combined the best features of direct drive with those of torque converter drive tractors.

D9 with a 100-foot Glencoe plow.

D9 in rough road conditions.

D9G with Holt special brush clearing attachment, 1963.

No. 9 ripper for the D9.

D9 with Kelley ripper for pipeline trenching.

Tandem D9G for pushloading application. "Quad-Track" arrangement permitted full utilization of the power of two 385 hp tractors.

D9G with Holt special brush clearing attachment. 1963.

D9G with Holt funnel dozer and root plow clearing brush on the King Ranch.

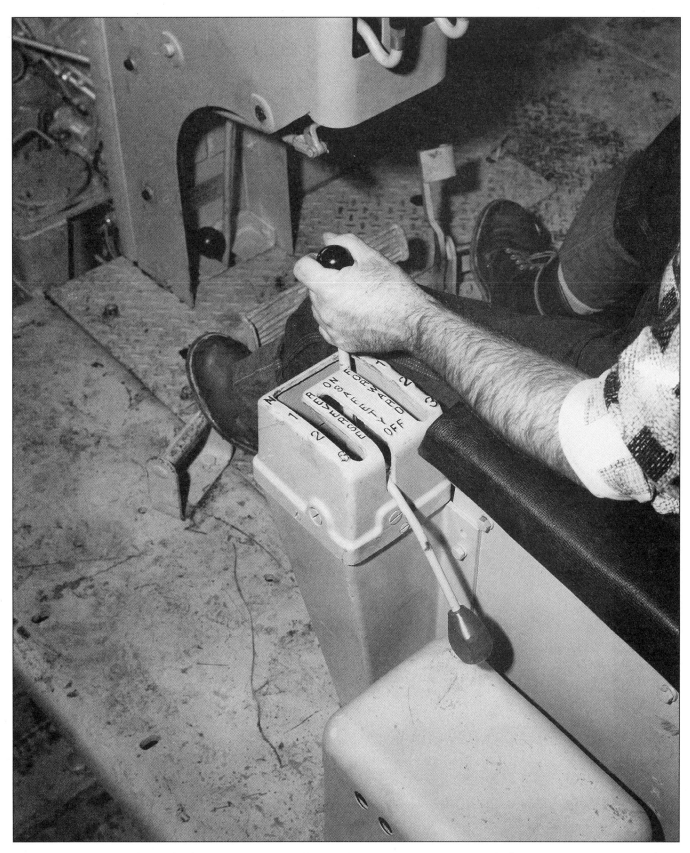
Horseshoe-shaped guide for gear range selector lever on D9G power shift transmission. The first leg controls for forward movement offered a range of low, intermediate, and high. Second leg offered the same range in reverse movement, with neutral position in the curve of the horseshoe. Rear lever was a manual safety used to lock the gear selector level in neutral.

D9G offered 385 hp and improved torque divider power shift transmission. Controlled turbocharging and aftercooling of intake air contributed to a 15% boost in horsepower.

INDEX

Best Autotractor, p. 66

Best "Humpback" Thirty, pp. 69-70

Best Sixty, pp. 75-83

Best Thirty, pp. 83-87

Best Tractor Engine, p. 66

Best 8-16 ("Pony"), p. 73

Best 16-25 (Model 25), pp. 73-75

Best 40, p. 71

Best 75, pp. 67-69

Best 90, p. 72

Caterpillar Diesel Fifty, pp. 165-166

Caterpillar Diesel Forty, pp. 180-181

Caterpillar Diesel Seventy-Five, pp. 177-180

Caterpillar Diesel Sixty, pp. 151-154

Caterpillar Diesel Sixty-Five, pp. 155-156

Caterpillar Diesel Thirty-Five, pp. 170-175

Caterpillar D2, pp. 188-201; 248

Caterpillar D4, pp. 202-218; 248; 264; 270

Caterpillar D6, pp. 218-227; 240

Caterpillar D7, pp. 228-243; 253; 261

Caterpillar D8, pp. 244-273; 274

Caterpillar D9, pp. 273-284

Caterpillar experimental machines, pp. 151; 273-275

Caterpillar Fifteen, pp. 143-151; 159

Caterpillar Fifty, pp. 161-164

Caterpillar R2, p. 187

Caterpillar R4, p. 214

Caterpillar R35 (Gasoline Thirty-Five), p. 168

Caterpillar RD4, pp. 203-205; 228

Caterpillar RD6, pp. 218; 220

Caterpillar RD7, pp. 229-231

Caterpillar RD8, p. 245

Caterpillar Sixty, pp. 100-112; 155

Caterpillar Sixty-Five, pp. 155-156

Caterpillar Ten, pp. 135; 137-143

Caterpillar Thirty (S and PS Series), pp. 113-124; 159

Caterpillar Thirty (6G Series), pp. 182-187

Caterpillar Thirty-Five, pp. 166-170

Caterpillar Twenty, pp. 124-133; 141

Caterpillar Twenty-Two, pp. 156-159

Caterpillar Twenty-Five, pp. 159-161

Caterpillar Twenty-Eight, p. 175

Caterpillar 2 Ton, pp. 91-100

Caterpillar 5 Ton, pp. 90-91

Caterpillar 10 Ton, p. 89

Holt experimental machines, pp. 7-9

Holt Midget, pp. 27-30

Holt steam crawlers, pp. 8; 10-11

Holt T-11, p. 33

Holt T-29, pp. 31-32; 40

Holt T-35, pp. 40; 57-61

Holt 2 Ton, pp. 39; 57-64

Holt 5 Ton, pp. 30-40; 64

Holt 10 Ton, pp. 41-57

Holt 12-18 (Model 18), pp. 23-25

Holt 20-30 ("Baby" 30), pp. 17-18

Holt 40, pp. 8-12

Holt 40-45, p. 13

Holt 45 (Standard), p. 14; (Military), pp. 25-27

Holt 60, p. 15

Holt 75, pp. 19-21

Holt 120, pp. 21-23

The Iconografix Photo Archive Series includes:

AMERICAN CULTURE
AMERICAN SERVICE STATIONS 1935-1943	ISBN 1-882256-27-1
COCA-COLA: A HISTORY IN PHOTOGRAPHS 1930-1969	ISBN 1-882256-46-8
COCA-COLA: ITS VEHICLES IN PHOTOGRAPHS 1930-1969	ISBN 1-882256-47-6
PHILLIPS 66 1945-1954	ISBN 1-882256-42-5

AUTOMOTIVE
FERRARI PININFARINA 1952-1996	ISBN 1-882256-65-4
GT40	ISBN 1-882256-64-6
IMPERIAL 1955-1963	ISBN 1-882256-22-0
IMPERIAL 1964-1968	ISBN 1-882256-23-9
LE MANS 1950: THE BRIGGS CUNNINGHAM CAMPAIGN	ISBN 1-882256-21-2
LINCOLN MOTOR CARS 1920-1942	ISBN 1-882256-57-3
LINCOLN MOTOR CARS 1946-1960	ISBN 1-882256-58-1
MG 1945-1964	ISBN 1-882256-52-2
MG 1965-1980	ISBN 1-882256-53-0
PACKARD MOTOR CARS 1935-1942	ISBN 1-882256-44-1
PACKARD MOTOR CARS 1946-1958	ISBN 1-882256-45-X
SEBRING 12-HOUR RACE 1970	ISBN 1-882256-20-4
STUDEBAKER 1933-1942	ISBN 1-882256-24-7
STUDEBAKER 1946-1958	ISBN 1-882256-25-5
VANDERBILT CUP RACE 1936 & 1937	ISBN 1-882256-66-2

TRACTORS AND CONSTRUCTION EQUIPMENT
CASE TRACTORS 1912-1959	ISBN 1-882256-32-8
CATERPILLAR MILITARY TRACTORS VOLUME 1	ISBN 1-882256-16-6
CATERPILLAR MILITARY TRACTORS VOLUME 2	ISBN 1-882256-17-4
CATERPILLAR SIXTY	ISBN 1-882256-05-0
CLETRAC AND OLIVER CRAWLERS	ISBN 1-882256-43-3
ERIE SHOVEL	ISBN 1-882256-69-7
FARMALL CUB	ISBN 1-882256-71-9
FARMALL F-SERIES	ISBN 1-882256-02-6
FARMALL MODEL H	ISBN 1-882256-03-4
FARMALL MODEL M	ISBN 1-882256-15-8
FARMALL REGULAR	ISBN 1-882256-14-X
FARMALL SUPER SERIES	ISBN 1-882256-49-2
FORDSON 1917-1928	ISBN 1-882256-33-6
HART-PARR	ISBN 1-882256-08-5
HOLT TRACTORS	ISBN 1-882256-10-7
INTERNATIONAL TRACTRACTOR	ISBN 1-882256-48-4
INTERNATIONAL TD CRAWLERS 1933-1962	ISBN 1-882256-72-7
JOHN DEERE MODEL A	ISBN 1-882256-12-3
JOHN DEERE MODEL B	ISBN 1-882256-01-8
JOHN DEERE MODEL D	ISBN 1-882256-00-X
JOHN DEERE 30 SERIES	ISBN 1-882256-13-1
MINNEAPOLIS-MOLINE U-SERIES	ISBN 1-882256-07-7
OLIVER TRACTORS	ISBN 1-882256-09-3
RUSSELL GRADERS	ISBN 1-882256-11-5
TWIN CITY TRACTOR	ISBN 1-882256-06-9

RAILWAYS
CHICAGO, ST. PAUL, MINNEAPOLIS & OMAHA RAILWAY 1880-1940	ISBN 1-882256-67-0
CHICAGO & NORTH WESTERN RAILWAY 1975-1995	ISBN 1-882256-76-X
GREAT NORTHERN RAILWAY 1945-1970	ISBN 1-882256-56-5
MILWAUKEE ROAD 1850-1960	ISBN 1-882256-61-1
SOO LINE 1975-1992	ISBN 1-882256-68-9
WISCONSIN CENTRAL LIMITED 1987-1996	ISBN 1-882256-75-1

TRUCKS
BEVERAGE TRUCKS 1910-1975	ISBN 1-882256-60-3
BROCKWAY TRUCKS 1948-1961*	ISBN 1-882256-55-7
DODGE TRUCKS 1929-1947	ISBN 1-882256-36-0
DODGE TRUCKS 1948-1960	ISBN 1-882256-37-9
LOGGING TRUCKS 1915-1970	ISBN 1-882256-59-X
MACK® MODEL AB*	ISBN 1-882256-18-2
MACK AP SUPER-DUTY TRUCKS 1926-1938*	ISBN 1-882256-54-9
MACK MODEL B 1953-1966 VOLUME 1*	ISBN 1-882256-19-0
MACK MODEL B 1953-1966 VOLUME 2*	ISBN 1-882256-34-4
MACK EB-EC-ED-EE-EF-EG-DE 1936-1951*	ISBN 1-882256-29-8
MACK EH-EJ-EM-EQ-ER-ES 1936-1950*	ISBN 1-882256-39-5
MACK FC-FCSW-NW 1936-1947*	ISBN 1-882256-28-X
MACK FG-FH-FJ-FK-FN-FP-FT-FW 1937-1950*	ISBN 1-882256-35-2
MACK LF-LH-LJ-LM-LT 1940-1956 *	ISBN 1-882256-38-7
MACK MODEL B FIRE TRUCKS 1954-1966*	ISBN 1-882256-62-X
MACK MODEL CF FIRE TRUCKS 1967-1981*	ISBN 1-882256-63-8
STUDEBAKER TRUCKS 1927-1940	ISBN 1-882256-40-9
STUDEBAKER TRUCKS 1941-1964	ISBN 1-882256-41-7

*This product is sold under license from Mack Trucks, Inc. All rights reserved.

The Iconografix Photo Album Series includes:
CORVETTE PROTOTYPES & SHOW CARS	ISBN 1-882256-77-8
LOLA RACE CARS 1962-1990	ISBN 1-882256-73-5
McLAREN RACE CARS 1965-1996	ISBN 1-882256-74-3

The Iconografix Photo Gallery Series includes:
CATERPILLAR	ISBN 1-882256-70-0
INTERNATIONAL HARVESTER TRACTORS	ISBN 1-882256-81-6 AVAILABLE 3/15/98

All Iconografix books are available from direct mail specialty book dealers and bookstores worldwide, or can be ordered from the publisher. For book trade and distribution information or to add your name to our mailing list contact

Iconografix
PO Box 446
Hudson, Wisconsin, 54016

Telephone: (715) 381-9755
(800) 289-3504 (USA)
Fax: (715) 381-9756

MORE GREAT BOOKS

CATERPILLAR SIXTY
Photo Archive
ISBN 1-882256-05-0

CATERPILLAR MILITARY TRACTORS VOL. 1 *Photo Archive*
ISBN 1-882256-16-6

CATERPILLAR MILITARY TRACTORS VOL. 2 *Photo Archive*
ISBN 1-882256-17-4

ERIE SHOVEL *Photo Archive*
ISBN 1-882256-69-7

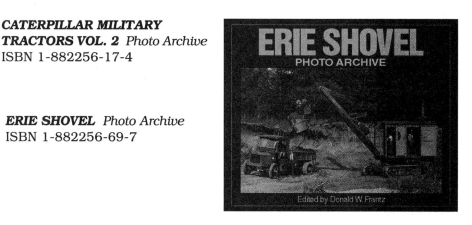

HOLT TRACTORS *Photo Archive*
ISBN 1-882256-10-7

INTERNATIONAL TD CRAWLERS 1933-1962 *Photo Archive*
ISBN 1-882256-72-7

CLETRAC & OLIVER CRAWLERS
Photo Archive ISBN 1-882256-43-3

RUSSELL GRADERS
Photo Archive ISBN 1-882256-11-5

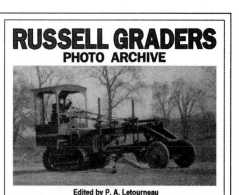